Induction watt-and varhour meters

1. How the induction watthour meter works 3
2. The voltage coil 6
3. The current coil 7
4. The disk 7
5. The braking magnet 7
6. The bearings 8
7. The meter constant 8
8. Adjustments of the watthour meter 9
8.1 Low load adjustment 9
8.2 High load adjustment 9
8.3 Power factor adjustment 10
9   One-phase electric energy reading 11
    9.1 Two-and-a-half element
    for one phase reading 13
10 Three-phase electric
    energy reading 15
    10.1 The three-element watt hour meter 15
    10.2 The two-element watt hour meter 17
    10.3 Two-and-one half element meter
    in delta systems 22

11 Varhour meters 25

11 Varhour readings using
    watthours meters 27

FOREWORD

The objective of Induction Watthour Meters is to give information about how to connect watthour meters to read one-phase and three-phase energy, individually or simultaneously, and how the most important watthour meters types work. This information applies to induction or also called electromechanical watthour meters, massively used in both one phase and three-phase electric energy reading. Contains basic information about the structure and work of the watthour meter.

The induction watthour meter has suffered changes since they first developed and patented, but the principle remains the same: the interaction of the magnetic lines of a potential or voltage coil and magnetic lines associated to the current flowing in the feeder. These magnetic lines act upon a disc that rotates coupled with a gear trend that registers the kilowatthours consumed by the load.

There are digital watthour meters, but the robust construction of the induction watthour meter holds its ground in the massive measurement of electric energy consumed by the customer in the household and in the industry. Usage of digital watthour meters limited so far.

So, the basic elements on the induction watthour meter are the potential or voltage coil, the current coil and a disk that rotates in the gap between the voltage and the current coil.

A few words have been devoted to the varhour meters, as it might be necessary to get information about the reactive component of the load

1. How the induction watthour meter works

The watthour meter is nothing but an induction motor. It works under the same principles of the induction motor: the reaction between a steady magnetic field and the magnetic field associated to an electric current flowing in a conductor. On one side, the magnetic lines associated to the current will add to the magnetic lines of the steady field. On the other side of the conductor the magnetic lines associated to the current will subtract from the lines of the steady field. This makes an effect, as it was a rubber band that tries to move the conductor to the position where the magnetic field is weaker. If the conductor is placed on a pivot or central axis, it will rotate.

In the case of the induction watthour meter the arrangements of the coils is such, that the interaction of the fields make the disk rotate at a speed *proportional to the consumed electric energy.*

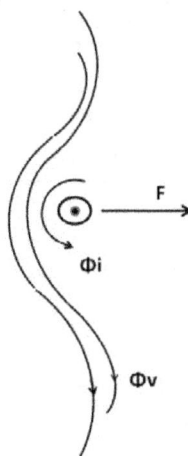

Fig 1.1 Interaction of a current and a steady magnetic field.

The dot in fig 1.1 shows that the current is flowing towards the paper. Using the right hand rule, we can determine that the magnetic lines associated to the current will flow counter clockwise. The magnetic field on the left side of the illustration will be stronger than on the right side and a force F will push the conductor to the right.

In the case of the watthour meter the conductor will be the disk of the meter and the eddy currents induced on the disk by the current flowing in the current coil will interact with the magnetic lines coming from the voltage coil.

Fig 1.2 shows a schematic arrangement of the induction watthour meter

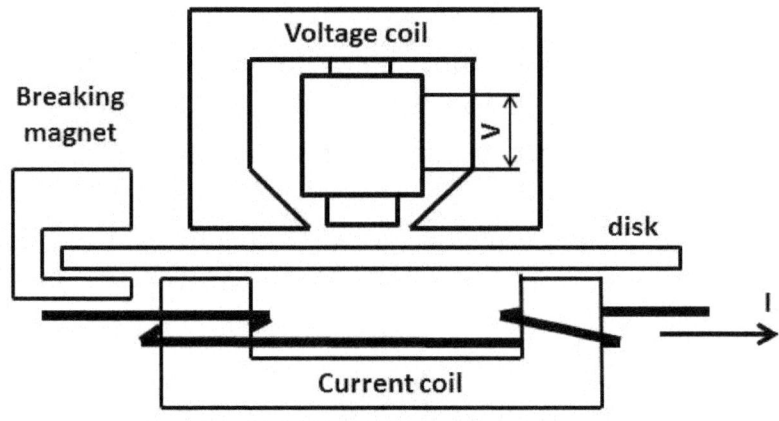

Fig 1.2 Schematic watthour meter

The meter will read correctly when the voltage and current flux and 90^0 with respect to one another.

The driving torque of the disk will be proportional to voltage magnetic flux multiplied by current magnetic flux, that is:

(1.1) $\qquad$ Torque = $\Phi_V \cdot \Phi_I \cdot \sin \alpha$

When angle is $90°$ sin α = 1.0 If α is not $90°$ but a different α angle, then the angle between the two flux will be $90° - \alpha$ and $\sin(90° - \alpha) = \cos \alpha$. This is important to note, that the disk will rotate proportional to the cosine of the angle, which is the power factor of the load. It means the disk will rotate proportional to the **cosine** of the angle between the two fluxes. The cosine defines the effective or active power demanded by the load to do useful work, therefore the disk will rotate proportional to the active or effective load.

Figure 1.3 shows that the disk will rotate proportional to the *projection* of the flux on the x axis. As the flux and load current are in phase, the projection of the current vector on the voltage applied to the meter will determine the energy registered by the meter.

If the projection of the current to which the flux is associated points in the same direction of voltage vector, the disk will rotate in the right direction. If the projection points opposite to the direction of the voltage vector, the rotation will reverse.

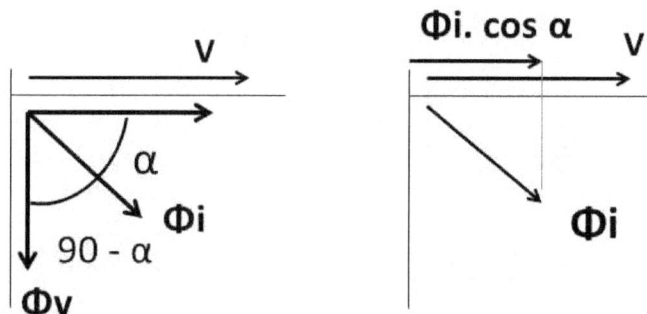

Fig. 1.3 Relationship between voltage and current flux if angle is different from $90°$

where $\qquad$ Φ_V is the magnetic flux originated by the voltage coil

Φ_I is the magnetic flux associate to the current

α is the angle between voltage and current flux different from 90^0

2. The voltage coil

The voltage coil is a coil composed of a large number of turns of a small cross section wire wound on a magnetic core. The voltage coil has two small gaps at each side to force the magnetic lines to cut through the disk. The vector representing the magnetic flux originated by this coil should be lagging the vector of the voltage connected to the coil.

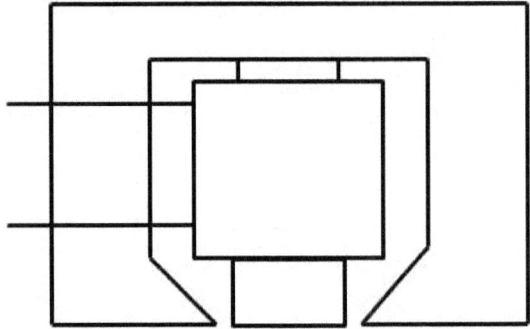

Fig 2.1 Schematic voltage coil in an induction watthour meter

Due to the great inductance of the voltage coil current will lag the voltage by 90^0. The magnetic lines or magnetic flux is in phase with the current and will lag the voltage as well.

3. The current coil

The current coil is made of a few turns a large cross section wire wound on a two-post core. The turns in both cores are wound in opposite directions forcing the magnetic flux associated to the current to cross the disk upward on one side and downwards on the other. The eddy currents originated by the current flux react with the voltage flux and originates the driving torque on the disk that makes it rotate.

Due to the small number of turns and the relatively large cross section of the wire on the current coil, the magnetic flux will be in phase with the current.

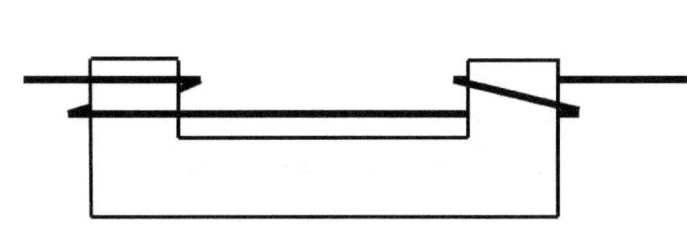

Fig 3.1 Schematic current coil in an induction watthour meter

4. The disk

The disk is usually made of aluminum to make it lighter and reduce friction on the lower bearing on which it stands. The upper bearing that keeps the disk in position is like a needle-like guide that keeps the disk upright.

5. The breaking magnet

If we allow the disk rotate freely it will do it at a speed close to 3,600 turns/minute, what is inconvenient for the design and wear of the system. A magnetic brake is used to reduce the speed of the disk and engage it to a mechanism that registers the energy consumed. The coupling between the disk shaft and the registry is accomplished by means of a worm gear. The braking magnet is shown on the left of fig 1.2. Note that the disk is rotating in the magnet gap.

6. The bearings

The lower bearing supports the weight of the disk. It is usually made of a hard, jewel-like substance. This bearing is designed to lower the friction of the disk. Some type of meters use magnetic bearings.

The piece of magnet placed at the bottom to support the disk has the same polarity as the piece of magnet placed on the shaft, in such a way, that the disk rotates like floating on the lower bearing. This arrangement is used specially in the multiple-element meters where there are several disks supported on the same shaft.

7. The meter constants

There are several constants in the meter. The meter constant itself known as Kh, represents the number of watthours equivalent to one turn. The gear ratio is the number of turns to make the first pointer make a full turn. The register ratio is the number of turns of the meshing wheel to make the first pointer go a full turn, but the most important constant is the Kh constant. The calibration of the meter is done based on the Kh of the meter.

The Kh of the meter is determined by design and is reflected on the nameplate of the meter. All adjustments and calibration of the meter are done to keep the meter to this value.

Knowing the Kh of the meter the power delivered instantly can be determined. If Kh represents the number of watthours per turn of the disk we can write:

(7.1) $\qquad Kh = wh/turn$

We don't want to wait hours to know the power of the feeder, we usually use a stop watch to determine time, so we have to convert the h to seconds and 7.1 will look his way:

(7.2) $\qquad K_h = W \cdot (sec/3600)/turn$

Therefore, if we take N turns to make sure that we can read the stop watch right we get for W on (7.2)

(7.3) $\qquad W = (3{,}600 \cdot K_h \cdot N)/\text{time in sec}$

If we want the instant load in kW we modify (7.3) and obtain

(7.4) $\qquad kW = (3.6 \cdot K_h \cdot N)/\text{time in sec}$

If we want to be more accurate we have to read the time the meter takes in one turn. If we take several turns we will have the average power taken during the N turns of the disk. Usually we take 5 or 10 turns.

8.0 Adjustments of the induction watthour meter

8.1 Low load adjustment.

We can reduce friction on the lower bearing, but we cannot fully eliminate it. For US standards low load is considered as 10% of full load. In order to reduce friction the practical meter introduces a compensation torque to help the disk cope with the low load and keep its accuracy.

The low load adjustment is an additional short circuited small coil placed under the voltage coil, somewhat shifted to one side. This introduces ar shifted flux that introduces an additional torque to help the disk deal with friction during low load. For load lower than 10% the meter cannot keep its accuracy, as 10% is the standard for low load calibration.

8.2 Full load adjustment

Full load adjustment is necessary because the meter parameters can vary with time, heat and frequency. The braking magnet can lose a bit of its breaking capacity due to weaker magnetic lines, etc. The full loac adjustment is done by shifting the position of the magnet on the disk. There is a worm screw that moves the position of the magnet inwards toward the center of the disk or outwards to the edge. The more the magnet moves inwards, the more magnetic lines go through the disk and the stronger the braking effect. After having changed the magnet's position there is a screw that fixes it position on the disk.

Other manufacturers use a shunt to bypass part of the magnetic lines making the breaking effect stronger or weaker.

8.3 Power factor adjustment

As we saw in 1.1 the torque of the induction meter is proportional to the product of the voltage-and current magnetic flux and the sine of the angle between them ($\Phi_V \times \Phi_I \times \sin\alpha$)

The sine curve is flat at the top, it means there is not much variation of the function for small variations in angle. The larger the angle α, the larger the reading error.

The magnetic flux originated by the voltage coil is $90°$ lagging the voltage and the magnetic flux associated to the current is in phase with the current. Due to the resistance mainly in the voltage coil the voltage flux will not lag the voltage exactly $90°$ as shown in fig. 8.3.1. Like in the case of the low load adjustment, we have to help the meter by introducing and additional magnetic flux that shifts the real flux back to its $90°$ position, so the meter can read correctly.

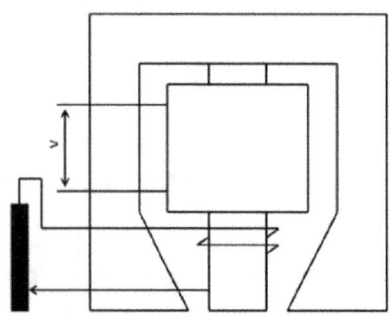
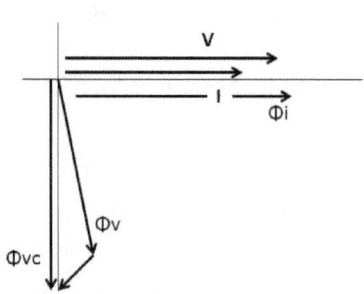

Fig 8.3.1 Power factor compensation on voltage coil. V: voltage, I: curren

Φ_V : voltage flux, Φ_{VC} corrected voltage flux

The additional magnetic flux is introduced be winding some turns on the post where the voltage coil is in series with a variable resistor that places the additional flux in the correct position respect current flux.

Some European manufacturers prefer to place the additional turns and the resistor on the current core and shit the current flux to meet the 90° requirement. The additional compensating flux pushes the current flux to its position at 90^0 to voltage flux, as shown in figure 8.3.2.

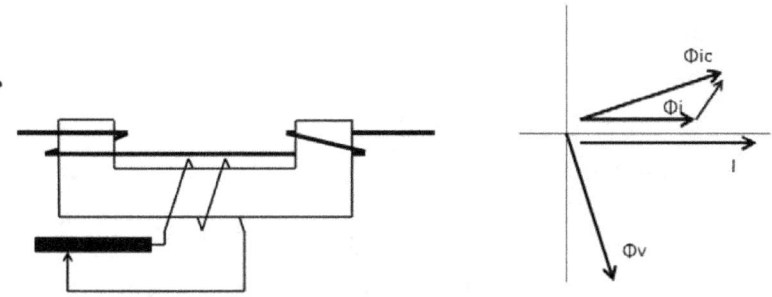

Fig 8.3.2 Power factor compensation on current coil. I: current, Φ_I current flux, Φ_{IC} corrected current flux

9.0 One phase electric energy reading

One-phase electric energy is read with one phase watthour meter. They have only one element. The combination of one voltage and one current coil working on one disk and the brake magnet with the rest of the adjusting devices is called an *element*.

We can suppose that three-phase electric energy requires one element for each phase and so it is. The disk of all the elements are attached to the sane shaft, the motion of the shaft is transferred to a geared registry by means of a worm gear that transfers the motion of all the disks to the gear train register and records the total energy of all the here phases. However, to make the watthour meter more cost effective the manufacturer tries to use a few elements as possible without compromising the accuracy in the reading.

In the following sections we will analyze he different options to measure electric energy with as few elements as possible. Every element has a breaking magnet and each element is calibrated individually for low load (10%), full load (100%) and power factor (0.5). The power factor calibration is required only at full load.

In Europe the calibration is three phase, the meter is calibrated as one unit and one element may be plus, the others may be minus so the combined calibration stays within the limits of tolerance specified in the standard. In North and Central America the calibration goes individually for each element and is performed as if it each element was a one-phase meter. The way of calibration and the calibration facilities may vary depending on the Company that supplies electric energy.

As energy is power multiplied by time, it is enough if we analyze the power expressions to determine how the watthour meter will read.

A one-phase transformer supplies one-phase load. Its secondary is split in two halves of 120 V each. The middle point is grounded. The customer receives one phase and the ground wire. Fig. 9.1 shows that the energy of each phase is read by one watthour meter per phase.

Supposing that $V_{AN} = V_{BN} = V_{AB}/2$ for load connected to V_{AN} WA will read

(9.1) $\qquad WA = (V_{AB}/2) \cdot I_A \cdot \cos \varphi_{an}$

The watthour meter in phase B will read

(9.2) $\qquad WB = (V_{AB}/2) \cdot I_B \cdot \cos \varphi_{bn}$

In order to determine the energy consumed by the load connected to V_{AB} we have to add (9.1) + (9.2).

(9.3) $\qquad WAB = (V_{AB})/2 \cdot I_{AB} \cdot \cos \varphi_{ab} + (V_{AB}/2) \cdot I_{AB} \cdot \cos \varphi_{ab} = V_{AB} \cdot I_{AB} \cdot \cos \varphi_{ab}$

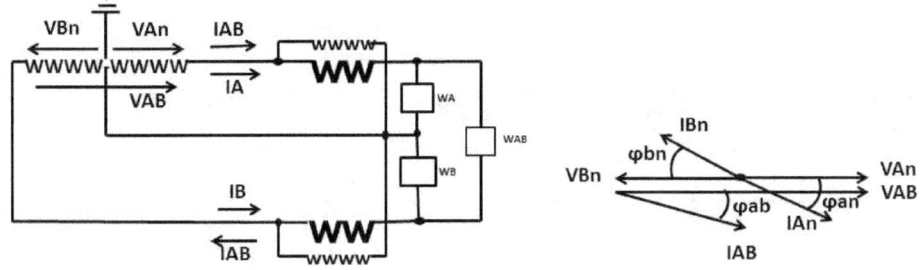

Fig 9.1 One phase energy reading with two watthour meters

In fig 9.1 note that the position of vector V_{BN} is opposite to V_{AB}, therefore reading at WB should be reversed, but note that current I_{AB} is coming into WB in opposite direction to polarity of current coil, it means that the current vector I_{AB} will reverse giving a positive reading. This a very important point when analyzing wattmeters and watthour meter connections: **reverse polarity makes current and/or voltage vector reverse.** Polarity shows to where the vector point. This is also valid in case of reading instant power with wattmeters.

Current I_{AB} returns through meter on phase B on the wrong polarity side, it will reverse the vector what comes handy because this will make the projection of the current in the same direction as V_{Bn} so the reading will be positive in (9.2).

9.1 The two and a half elements for one phase readings.

We have analyzed how phase to neutral load and phase to phase load is read with two watthour meters. It means that we would have to use two watthour meters in case the customer had load between phase and ground (neutral wire) and between the two phases.

This can situation be solved using only one watthour meter with two current coils, one for phase A and one for phase B., and one voltage coil connected to V_{AB} as shown schematically in fig 9.1.1.

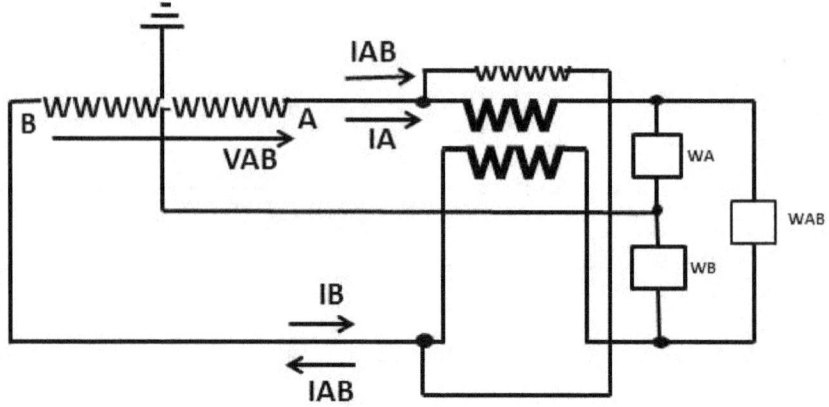

Fig 9.1.1 Two phase readings with one meter

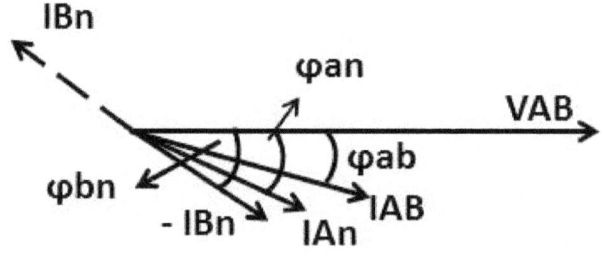

Fig 9.1.2 Vector diagram for scheme shown in figure 9.1.1

For this type of meter we use only voltage V_{AB}. The condition for the correct work of this arrangement is:

1) each current coils must have ½ of the turns

2) current coil in phase B must be wound in reverse of that of phase A.

The calibration of the type of meter requires that both current coils be connected in series during calibration to make one full coil composed of the two half-coils. As the calibration current returns in the wrong side of coil B, and this coil is wound backwards, the effect will be like the current vector cast a positive projection on the voltage vector. Wrong direction + backward winding = positive rotation.

Current I_{AB} will return in the wrong direction, but reverse turns will make a positive reading on. This will work right for current I_B too. This current needs to be reversed because now we are using V_{AB} for I_B, that is in opposite position to V_{AN}.

This type of meter will read for energy in phase AB (WAB):

(9.1.1) $\quad WAB = V_{AB}.(I_{AB}/2).\cos \varphi_{ab} + V_{AB}.(I_{AB}/2).\cos \varphi_{ab} = V_{SB}.I_{AB}.\cos \varphi_{ab}$

For energy on phase AN (WAn):

(9.1.2) $\quad W_{AN} = V_{AB}.(I_{AN}/2).\cos \varphi_{an} = V_{AN}.I_{AN}.\cos \varphi_{an}$

For energy on phase BN (WBn):

(9.1.3) $\quad W_{BN} = V_{AB}.(I_{BN}/2).\cos \varphi_{bn} = V_{BN}.I_{BN}.\cos \varphi_{bn}$

because $\quad V_{AB}.(I_{AN}/2) = (V_{AB}/2).I_{AN}$

10. Three phase electric energy reading
10.1 Three-element watthour meter.

For three-phase energy metering three-element, two-element, or two-and-a half element meters are used. The three-element meter is used to read electric energy in a wye system with four wires, three phases and the ground wire. Every element of the meter reads both one phase-and three-phase energy in each phase individually. The three disks of the three elements are stacked on the same shaft and all the read energy is transmitted to a common gear that is engaged in a worm gear on the shaft. Fig. 10.1 shows a schematic arrangement of the three-element meter in a wye system. In this case the total energy read by the meter will be given by:

(10.1) $\quad W_{ABC} = V_{AN} \cdot I_A \cdot \cos \varphi_A + V_{BN} \cdot I_B \cdot \cos \varphi_B + V_C \cdot I_C \cdot \cos \varphi_C$

For one phase energy supply phase A, B and neutral wire are used with one-phase transformer as we have seen before. For three phase energy metering load on phase C has to be also considered.

Only three-phase load is shown in figure 10.1.1a to avoid clutter that may make the diagram unclear. One phase load can be connected in one or all phases and the reading will include all energy, three-phase and one-phase.

Fig. 10.1.1b Shows the vector relationships for this type of connection

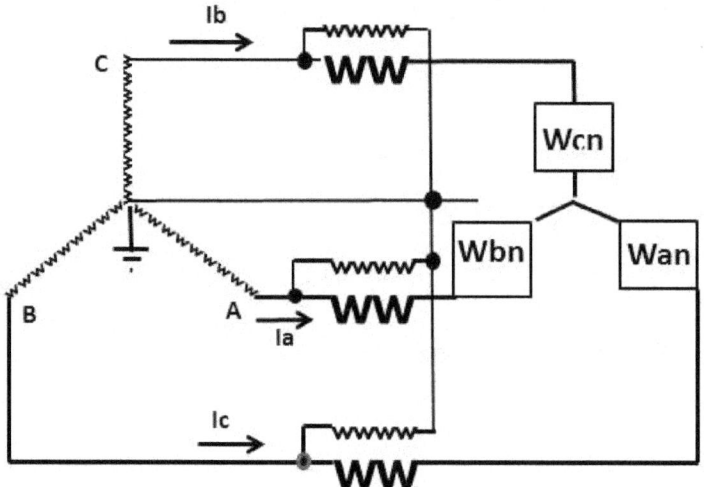

Fig 10.1.1a Three element what-hour meter

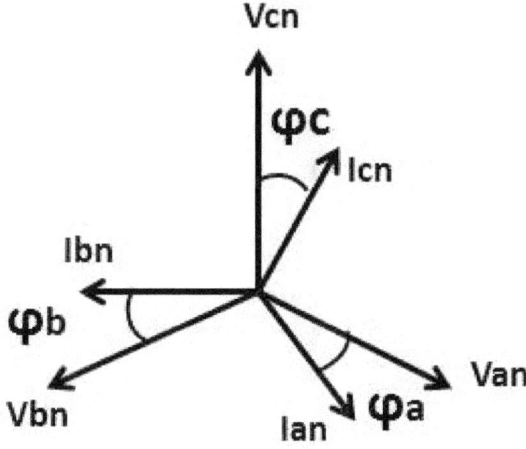

Fig 10.1.1b Three element meter vector diagram

10.2 The two element watthour meter

The two-element watthour meter is used to read balanced three phase energy only without one phase load. Two elements are used in this case connected as shown in fig 10.2.1a.

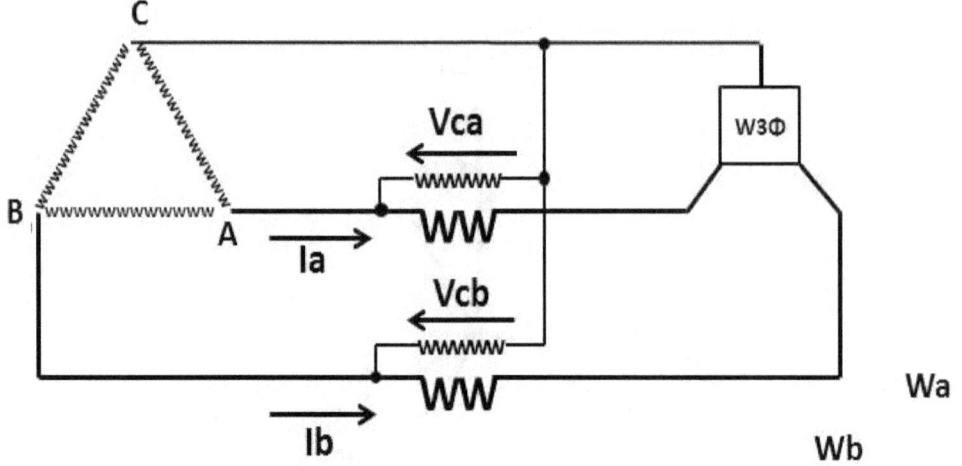

Fig 10.2.1a. Three phase reading with two element meter

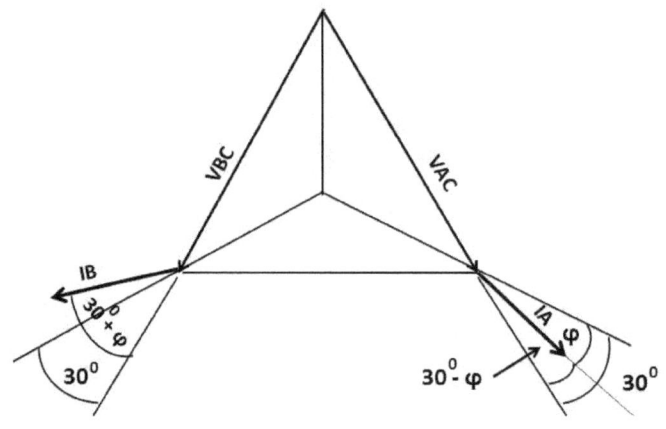

Fig 10.2.2b Vector diagram for two element meter

The three phase load is schematically shown in figure 10.2.1a as a wye. It can be either a wye or triangle connection in real life.

Figure 2.2.1b shows the vector diagram for this connection.

In the triangle or delta connection the power factor 1.0 reference is the wye embedded in the delta. When the load power factor (cos φ) is 1.0 on the line, the position if the current in each phase will be in phase with the wye inside the triangle (see *Transformer Connections* by same author).

In this case V_{BC} and V_{AC} have been drawn according to the polarity connection of voltage coil between phase C and B.

The energy read by the AC element will be:

(10.2.1) $Wh_{AB} = V_{AC} \cdot I_A \cdot \cos(30° - φ) + V_{BC} \cdot I_B \cdot \cos(30° + φ)$

Using the expression $\cos(a + b) = \cos a \cdot \cos b - \sin a \sin b$

$\cos(a - b) = \cos a \cdot \cos b + \sin a \cdot \sin b$

Assuming we have a balanced three-phase load we can write:

$$V_{AC} = V_{BC} = V_L$$

$$I_A = I_B = I_L$$

The load read by the meter on phase A:

(10.2.2) $Wh_A = V_L \cdot I_L \cdot \cos 30° \cdot V_L \cdot I_L \cdot \cos φ + V_L \cdot I_L \cdot \sin 30° \cdot V_L \cdot I_L \cdot \sin φ$

The load read by the meter in phase B:

(10.2.3) $Wh_B = V_L \cdot I_L \cdot \cos 30° \cdot V_L \cdot I_L \cdot \cos φ - V_L \cdot I_L \cdot \sin 30° \cdot V_L \cdot I_L \cdot \sin φ$

If we add (10.2.2) + (10.2.3) the terms containing the sinus function cancel themselves because they have opposite signs and we have left:

(10.2.4) $Wh_{AB} = 2 \cdot V_L \cdot I_L \cdot (\sqrt{3}/2) \cdot \cos φ = \sqrt{3} \cdot V_L \cdot I_L \cdot \cos φ$

That is the expression for the three phase active load. As energy is power multiplied by time, this result applies to watt-meter-and watthour meters connections as well.

Three phase load in a wye system can be also read with a two element meter and three current coils. In this case two phase voltages are used, the current of the third phase flows through the third current coil that is common for both elements.

Figure 10.2.3a shows the schematic construction of this type of meter. The condition for this meter to work right is that the load must be three phase and balanced. This meter will read right the three-phase load and one phase load connected to phase voltage, that is, between the phase and the ground wire. It will not read right one phase load connected to line voltage, so it is designed to read exclusively three phase load.

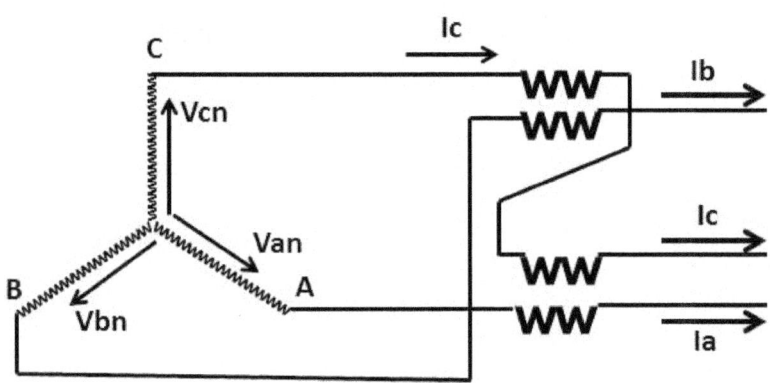

Fig 10.2.3a Schematic energy reading of three-phase load with two element wye meter.

Unlike the two-coil meter for one phase connection in Fig. 9.2, the number of turns is the same in all the three coils, considering the two coils for phase C as one coil.

Fig 10.2.3b shows the vector relationships that make this arrangement read right for three-phase load.

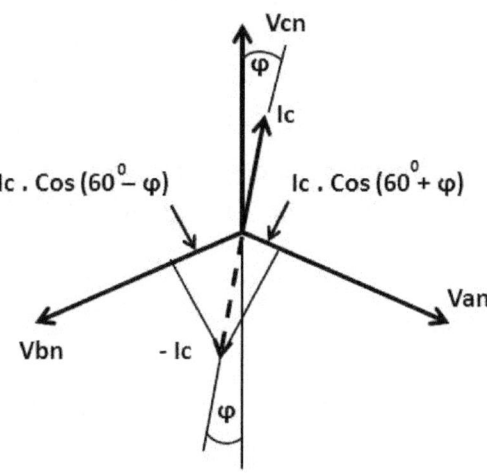

Fig 10.2.3b Vector diagram for the two coils wye meter

Note that the reverse polarity of the third common coil for both elements A and B makes vector I_C reverse, giving a positive projection on both V_a and V_b.

Reading in element B will be:

(10.2.4) $V_{BN} \cdot I_c \cdot \cos(60° - \varphi) = V_{BN} \cdot I_c \cdot \cos 60° \cdot \cos\varphi + V_{BN} \cdot \sin 60° \cdot \sin\varphi$

Reading on element A will be:

(10.2.5) $V_{AN} \cdot I_c \cdot \cos(60° + \varphi) = V_{AN} \cdot I_c \cdot \cos 60° \cdot \cos\varphi - V_{AN} \cdot I_c \cdot \sin 60° \cdot \sin\varphi$

This type of meter is also called two-and-a-half element meter because the voltage coil of the third element is missing.

The sine terms cancel and what is left is:

(10.2.6) $2 \cdot (V_N \cdot I_c)/2 \cdot \cos\varphi = V_N \cdot I_c \cdot \cos\varphi$

Assuming that the load is symmetrical, the reading for each of the other two-phase will also be: $V_N \cdot I \cdot \cos\varphi$. The total reading for all the three phases will be $3 \cdot V_N \cdot I \cdot \cos\varphi$. This the expected result for the three-phase load in a wye symmetrical system.

One phase load connected to line voltage will not be read right by this arrangement, so it will work for symmetric three-phase load. Following the principles exposed so far about how the vector is affected by polarity and taking the component of the current projected to the voltage, the reader can experiment finding the vector relationships in case one phase load would be connected, for example, between phase C and phase B.

10.3 Two and a half element meter to read one-and three phase load in a delta of triangle system.

The two-and-a-half element meter is used to read one phase and three phase load in the grounded delta used to supply one-and three phase load has one element that is basically the same as the one discussed in section 9 an shown in figure 9.2. A second element has been added to take phase C current. In this case one voltage coil for the AB element is missing, that's why this device is called *two-and-a-half element meter*.

The voltage coil of the new element is connected between phase C and the neutral or ground wire of the delta, as shown in figure 10.3.1a

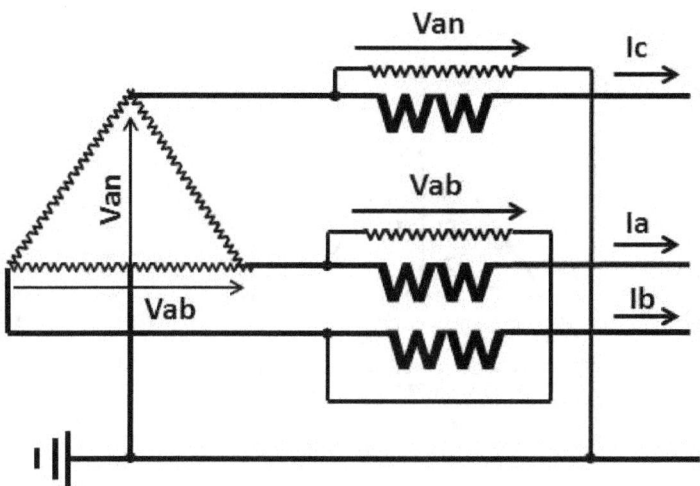

Fig. 10.3.1a Schematic arrangement of the two and half element meter for triangle connection

As we see, a new element has been added to the arrangement to read one-phase energy. Both current coils A and B have been wound with half the turns

The C current coil is wound with full number of turns and voltage V_{CN} is connected to the voltage coil.

We already discussed the energy reading in one-phase systems, so will skip this step this time and analyze what read each element only for three-phase load.

Figure 10.3.1b shows the vector relationships for all the elements. Note that current coil B is reversed, so the I_B vector of the three-phase load will reverse to cast a positive projection on V_{AB}.

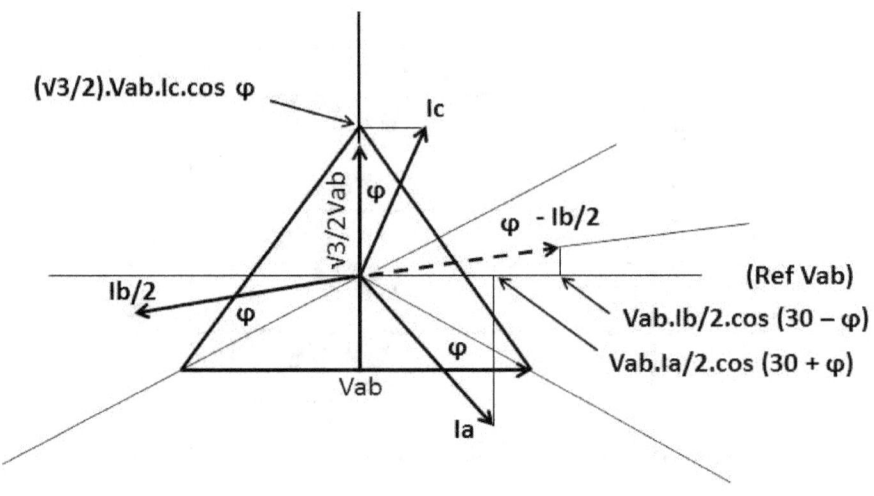

Fig 10.3.1b Voltage and current vector relationships in a two-and-a-half element meter for delta connection.

The element on phase C will read $(\sqrt{3}/2).V_{AB}.I.\cos\varphi$ because $V_{CN} = (\sqrt{3}/2).V_{AB}$.

Assuming that the three phase load is symmetrical and $I_A = I_B = I_C$ and $V_{AB} = V_{BC} = V_{AC} = V_L$ the element in phase A will read:

(10.3.1) $V_L.I/2.\cos(30-\varphi) = V_L.I/2.\cos 30.\cos\varphi + V_L.I/2.\sin 30.\sin\varphi$

The element in phase B will read:

(10.3.2) $V_L.I/2.\cos(30+\varphi) = V_L I/2.\cos 30.\cos\varphi - V_L.I/2.\sin 30.\sin\varphi$

The current vectors are divided by 2 because the currents coils in A and B are wound with half the turns required by the element. The current coil in phase C has the complete number of turns.

If we add (10.3.1) + (10.3.2) the terms containing the sin function cancel each other and we have:

(10.3.3) $2 \cdot (\sqrt{3}/2) \cdot V_L \cdot I/2 \cdot \cos\varphi = (\sqrt{3}/2) \cdot V_L \cdot I \cdot \cos\varphi$

We still have to add the reading for phase C. In this case we have:

(10.3.4) $\sqrt{3}/2 \cdot V_L \cdot I \cdot \cos\varphi + \sqrt{3}/2 \cdot V_L \cdot I \cdot \cos\varphi = \sqrt{3} \cdot V_L \cdot I \cdot \cos\varphi$ that is the expression for the three phase effective power, that is, the effective energy consumed by the three phase load.

11. Varhour meters

The alternating electric current has two components: one active or effective, that is the component that produces useful work and one useless, but necessary to sustain the magnetic fields that make the induction devices work. The active or effective component is measured by the watthour meter. Sometimes we might be interested in knowing the reactive component, as both components load the feeder.

We cannot say that the varhour meter measures reactive energy, as the term energy is related to useful work and the reactive component of the current does not do any useful work, but the varhour reading gives us an insight of the how the supplied power is used.

Having the energy readings and the period of time between them we can have an average value of the active and reactive component. The varhour reading divided by the watthour reading gives the average *tangent* of the angle between voltage and current.

Using trigonometric tables or some suitable trigonometric expression the cosine of the angle can be determine. This cosine is called the power factor of the load and is the ratio of the useful load to the total load.

Cosine function varies between 0 and 1,0, therefore, the power factor will vary between those values. The closer the power factor to 1.0, the more efficiently the supplied energy has been used (see *Reactive Power Management* by same author).

We have seen in previous sections that active or effective component of the current is the projection of the current over the voltage, that is, the component that is *in phase with* the voltage.

Reactive component of the current is 90^0 *lagging* the active component. If we shift 90^0 the position of the voltage used to read power or energy, the projection of the current over the shifted voltage will be in phase with the reactive component of the load.

Figure 11.1 shows this trick, as the meter does not know that he is reading something other than active energy.

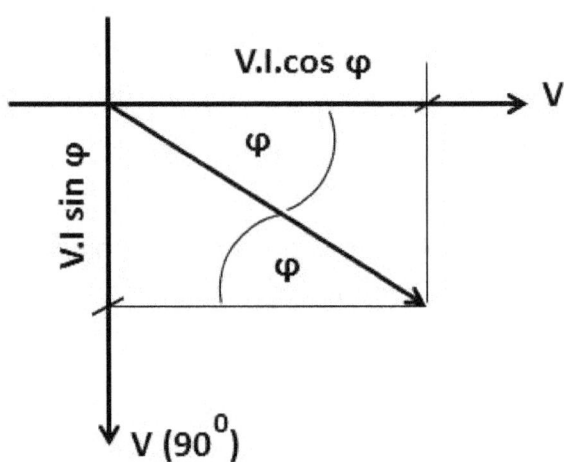

Fig. 11.1 Projection of the current on X axis to read active power, on Y axis to read reactive power

So the trick is to position the voltage connected to the meter 90^0 lagging perpendicular to the natural position of the system voltage. IN this case the meter will start rotating proportional to the *reactive* component, not the *active* component of the current.

12 Varhour readings using watthour meters.

If we can find a way to shift 90^0 the system voltage, we can improvise reactive power or energy readings with watthour meters. Take, for example, reactive readings with one active meter reading in a triangle or delta system, as shown in figure 12.1 we can take current I_C and voltage V_{AB} to do the reading.

In this case the rotation of the disk will be proportional to the *reactive* component of the current.

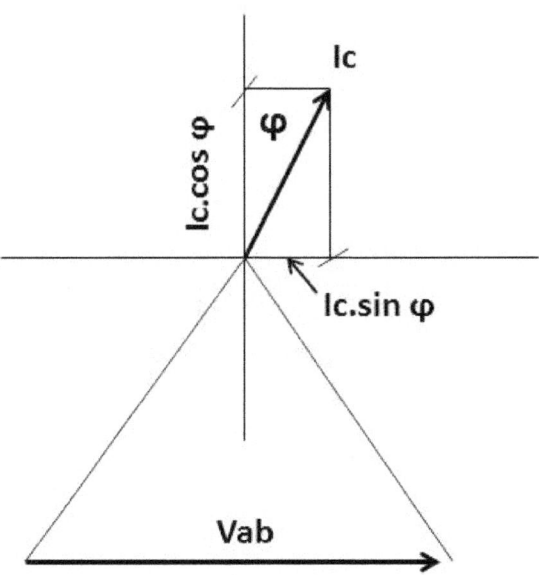

Fig 12.1 Connection to read reactive power using one watthour meter

In this specific case we are using one watthour meter assuming that the load is symmetrical. If this is the case, we have to multiply the reading by √3 to get an estimate of the three phase reactive reading.

Same procedure could be applied if the system was wye. In this case the voltage needed is phase voltage V_{CN} and we will use V_{AB} which is √3 higher than the required voltage. The energy for one phase will be $1/√3.V_{AB}.I.\cos φ$.

The total load would be $3/\sqrt{3}.V_{AB}.I.\cos \varphi = V_{AB}.I.\cos \varphi$

Using $3/\sqrt{3}$ as a factor for the reading a close estimate of the real varhour magnitude can be attained.

Let's take the vector diagram shown in figure 10.2.2b and do the same analysis with the voltage vectors shifted 90^0 lagging behind V_{AC} and V_{BC} as shown in figure 12.2.

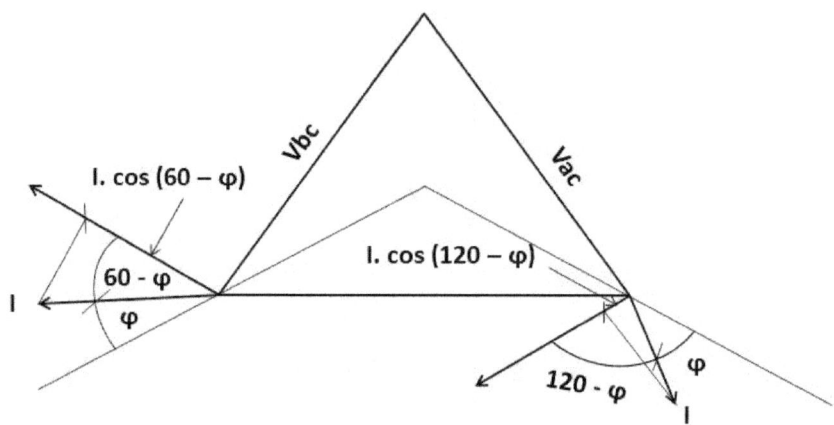

Fig. 12.2 Shifted vectors to read reactive power or "energy" in a Three phase delta connection.

The B element will read:

(12.1) $V90.I.\cos (60 – \varphi) = V90.I.\cos 60.\cos \varphi + V90.I.\sin 60.\sin \varphi$

The A element will read

(12.2) $V90.I. \cos (120 – \varphi) = V90.I.\cos 120. \cos \varphi + V90.I.\sin120. \sin \varphi$

$\cos 120^0 = - \cos 60^0$

$\sin 120^0 = \sin 60^0$

(12.2) can be modified this way:

(12.3) $- V90.I. \cos 60°.\cos \varphi + V90.I. \sin 60° . \sin \varphi$

Adding (12.3) + (12.2) the terms containing cosine function cancel themselves and we get:

(12.4) $2.u\ V90.I.\sin 60°.\text{sen } \varphi = \sqrt{3} .V90.I.\sin \varphi$

This is the expression for three phase reactive power and/or "energy".

The average power factor can be determined based on the watthour and varhour readings using the following expression:

(12.5) $\cos \varphi = 1/\sqrt{(1 + \tan^2 \varphi)}$

varhours/watthours = $\tan \varphi$, therefore

(12.6) $\cos \varphi = 1/\sqrt{[1 +(\text{var-hors/watthours})^2]}$

How can we shift the voltage in a practical way? If we add a resistor in series with the voltage coil we shift the current vector too. The angle between current and voltage will not be $90°$ anymore. There will be a voltage component on the resistor and another one on the inductance of the coil, as shown in figure 12.3.

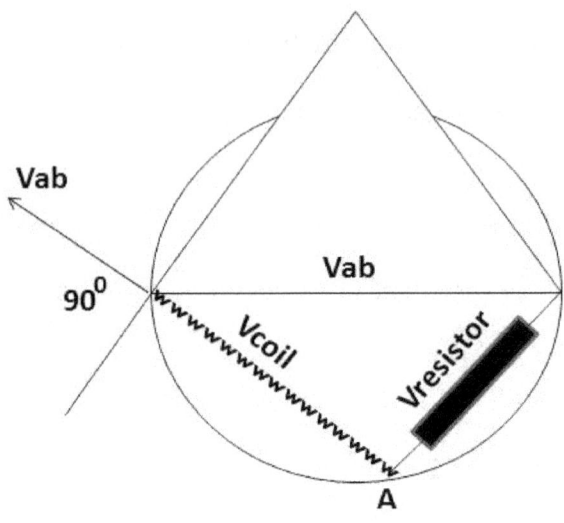

Fig. 12.3. Voltage shift caused by resistor in series with voltage coil.

By changing the value in ohm of the resistor point B can be shifted anywhere along the circle. If we make current B in element AB with the resistor shift, we will get a similar situation as in figure 12.2. A proper combination of shifted voltages and phase currents makes the trick.

The designers of varhour meters that are built to read only varhour take into consideration the proper number of turns of the voltage coil, cross section of the wire and adjustments to make the instrument work properly on the system to which it was intended.

Phase shifting can be obtained also using phase shifting transformers. These are transformers designed with taps and the proper compensation to provide 90^0 shifting. Figure 12.3 shows how phase shifting transformers are connected schematically in a delta system.

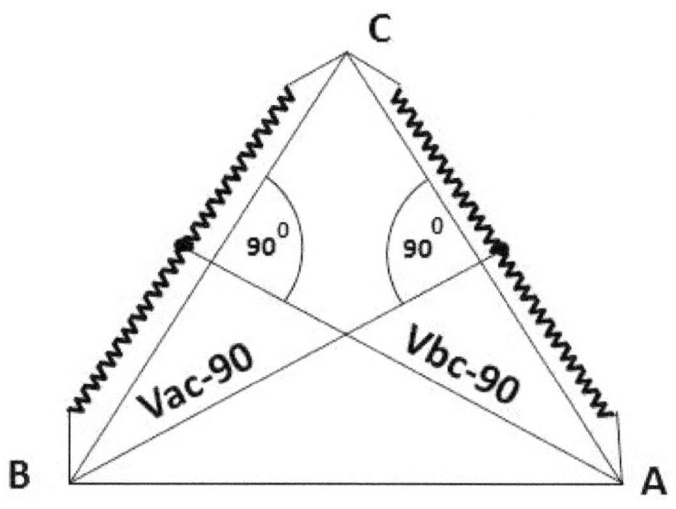

Fig 12.3 Phase shifting transformers in a delta system.

These phase shifting transformer are designed to be used with original watthour meters. Introduction of some multiplying factor could be necessary to obtain accurate war-or varhour readings.

We have to bear in mind that the voltage or current vector point to where the polarity of the coil is.

The analysis discussed so far are also valid for wattmeter or varmeter connections.

You can leave your opinion and/or suggestion in a review, at http://reactivepower.blogspot.com or write to rafbarr45@yahoo.com. If you found it useful, you can recommend this book to others that might be interested in the same topic.

www.ingramcontent.com/pod-product-compliance
Lightning Source LLC
Chambersburg PA
CBHW070730180526
45167CB00004B/1698